I0087450

WELCOME!

So you finally made the move. Maybe this move is your first move out to the country. Maybe you are going back to where you grew up or to grandpa's old property. Maybe you've spent the last 25 years in the rat race and you are ready to kick back on the rural route.

I don't know how you got here, but you are here now, somewhere past the city limit sign, so WELCOME!

That said, our quiet farms and rural properties don't take care of themselves. If you don't believe me, you will find out soon enough. To keep a property maintained, you either need a ton of cash available to pay someone to do it for you, OR, you need the time and the tools to do it yourself. Having the right tools can save lots of time.

You shouldn't fear the tasks you are about to take on. Being close to the land is ingrained in all of us. There are few jobs I can think of that are more rewarding than maintaining a piece of the earth. Running farm equipment, raising livestock, growing things, or simply managing the lawn creates a new set of challenges that should be embraced, not feared. The goal of this manual is to help you learn about the tools that can help you the most, but first, let me finish this pep talk:

As you read and make your transition to rural life I want you to keep two things in mind. 1) You are not the first person to make this transition. 2) There are folks out there who will be anxious to share their knowledge with you (even while we laugh and talk about you behind your back down at the bar on Saturday nights....just kidding.....sort of).

This manual will help you through the purchase of what is essentially the power plant of any farm (after brains and cash flow of course), the tractor. The goal is to teach you the basic features and functions of tractors in order to help you determine the machine that best suits your needs. The various options and configurations may seem vast at first, but most folks in your situation fall into only a handful of categories when it comes to what they actually need. My goal is to help you find the category that fits you best.

So again, congratulations! You've done what many only dream about. The fun part is just beginning!

Chapter 1: What is a Tractor?

This is not a dumb question. Read on.

Lets start on the most basic level: What IS a tractor? Many folks with no farm experience think of tractors simply as something that farmers drive. While this is true, a tractor is so much more than a slow mode of transportation.

To put it simply, the tractor is the power plant of the farm. A tractor is the top tool in the tool hierarchy. The tractor will power almost every piece of machinery one can imagine. The tractor will move anything that has wheels and some things that do not have wheels. Whatever task you wish to do, there is a tool that can be attached to the tractor to perform the task. Whether you have the smallest 25 horsepower sub compact, or a 250 horsepower row crop tractor, there are implements (various tools that attach to the tractor to perform specific tasks) designed to fit your tractor. Think of a Swiss Army Knife. The red oval shaped object looks simple enough, but there are dozens of things attached to it that are designed to do different jobs. Another more modern example would be the Dremel Tool. If you want to sand, you put on a sanding drum. If you want to polish something, you put on the polishing tool.

Without implements, a tractor is nothing more than a very slow mode of transportation. Without a tractor, implements are nothing more than heavy, dust-collecting objects to trip over.

(Tractor with a front-end loader. The loader is one type of implement we will discuss later)

The key to tractor buying is having a basic idea of what you wish to do with the tractor and to what degree you want to do it before you start shopping. If you have no idea, bare with me as I fill your head with ideas.

Definitions: **Tractor:** The power plant of the farm. The engine that provides propulsion and power to your implements.

Implement: Anything you attach to the tractor in order to perform a specific task. Most implements are made for a specific task. Examples include, rear blades for grading, a loader for lifting, a posthole digger for drilling holes, and a hay baler for baling hay.

Take a few moments and jot down what you know you plan to do with the tractor. Now would be a good time to walk around your property and think of everything you envision yourself doing. If you have a seasoned farm neighbor nearby, ask them to help you determine what you will need to do to manage your property. Walk the property and tell them your ideas. If you have NO clue as to what you will do with your tractor, keep reading, because on a farm there is ALWAYS SOMETHING TO DO!

Chapter 2: Basic Tractor Jobs

When you buy a piece of property, you will most likely start doing work around the house and the barn first. This is the area you most likely live in and it is the area you will probably want to get up and running first. See those overgrown shrubs? Your tractor can help you yank them out. Do you have old saggy paver stones or a busted up sidewalk? A tractor can help you excavate those areas. Is your "yard" totally overgrown with grass and weeds that the previous owner couldn't maintain anymore? Your tractor is there to help. Need a lift? A tractor with a loader is an indispensable tool.

Now look out past your yard. Do your fields need to be trimmed up? Do you need to build or repair fences? Are there trails you would like to clear for riding, hiking, hunting or biking? Would you like to put a duck blind out on the edge of the pond? Would you like to till a garden this spring? Do your fields need overseeding to feed livestock? Would you like to plow some ground and plant your own crops? Would you like to make your own hay? Would you like to cut, split, and haul firewood from your wooded areas? Do you expect snow this winter? Does your horseback riding ring need grooming? Do you need a riding ring built? Is your driveway a rutted bumpy mess?

If you answered "yes" to any of these questions, a tractor will help you immensely. Not only will your tractor help you, your tractor makes it possible for one person to do the tasks that once took dozens of man-hours. The tractor is a time saver, a money saver, and a back saver. Are you starting to understand why I refer to it as the power plant of the farm?

If this all sounds overwhelming, don't worry, it isn't. Besides, no one said you have to do all of the things listed in order to have a happy rural lifestyle. You many decide you only need one small tractor for things that are close to the homestead or you may want a larger tractor that is more suitable for what many will refer to as "field applications" (plowing, planting, baling hay, etc.).

Again, don't be overwhelmed. Remember your first computer? You probably bought it with one or two specific tasks in mind such as word processing and accounting. As time went on you began using the computer for more and more as you learned what the computer could do for you. Buying a tractor is very similar. You will initially buy it because you know you need it for one or two blatantly obvious tasks, but the longer you own the tractor, the more you will do with it.

Chapter 3: What Size Tractor do I need?

So, you have decided you need a tractor. Now, the question remains, what size?

First and foremost, lets focus on horsepower. Tractors are measured in horsepower. You will notice TWO horsepower numbers thrown around as you study literature and talk to folks about tractors: You will hear the terms "engine horsepower" and "PTO horsepower." Engine horsepower is the horsepower generated by the tractor's engine. PTO horsepower, for lack of time for a more technical definition, is the horsepower that is available to run implements. PTO horsepower is the CRITICAL number. Do not believe anyone who tells you otherwise. Any salesman who tries to sell a tractor based on engine horsepower does not know what he is talking about and does not deserve your business. Leave their store immediately!

Definition Time! **PTO: Power Take Off.** On the back of most tractors you will find a Power Take Off shaft. This shaft connects to various implements and draws power from the tractor's engine in order to run the implements.

(Power take off: PTO-The splined shaft in the center of the tractor's rear. This is used to power various implements)

So, how much horsepower do you need? Well, it all depends on what you want to do. If you are serious about baling hay, mowing large fields, and not outgrowing your tractor too fast, I recommend a tractor with a minimum of 65 PTO horsepower. Be careful, as there are some compact utility tractors rated at 60-65 ENGINE horsepower, but these tractors will not have nearly enough weight or PTO HORSEPOWER necessary to do many of the jobs you are envisioning.

If you only have a small piece of property and only wish to move dirt occasionally, cut the yard, and maybe do some snow removal, a 35 to 40 horsepower compact utility tractor will more than likely serve you well.

My advice to most folks is to do the following (and no, it isn't because I'm an up-selling salesman, it is because I don't want unhappy customers trading in machines after 12 months and hating me when they realize the depreciation hit they take in the first year):

Buy the largest tractor you can afford that meets any size constraints on your property (gates, barns, turn around areas). If you only need 65 HP but you can step up to a 75 or 85HP machine do it. The worst thing you can do is buy a tractor that is too small for your needs.

My reasons for encouraging folks to buy the largest tractor they can afford are as follows:

1) like all new items, you take your biggest hit in value the second you take them home.

2) Doing too much with a tractor that is too small is not only hard on the tractor, but it can be dangerous.

3) Once you find out how much you can do with a tractor, you will find yourself doing jobs you never thought you would do around your property and maybe even on the properties of others. That said, it is better to buy something a little larger than you may have intended and have a machine that fits you for a long time than to pay a lot for a tractor that you'll outgrow.

Most folks get scared when they see the tractor I recommend based on the property and job descriptions they give me.

"WHOA! That is WAY bigger than I envisioned!" they say, fear pouring from their face.

There is nothing to be afraid of. You will need someone to teach you how to operate the tractor, but if you learned to drive a car, you can learn to operate a tractor and implements. A good salesman will take the time to teach you. Neighbors with machinery experience will also be anxious to help you and to play with your new toy. So do not be intimidated, just keep an open mind and ask questions. We all have to start somewhere. The key is getting started!

Chapter Four: The Power Take Off (PTO), Expanded

Definition: PTO: Power Take Off. The mechanism on the tractor that takes power from the tractor's engine and delivers it to the implement.

A tractor is nothing more than a slow mode of transportation without implements. Many implements are useless without power. How does an implement, like a rear rotary cutter (also called bush hog), get power? Rotary cutters don't have their own engine, they don't run off of electricity, and they don't have hamsters running on wheels to power them. They have a hitch, a gearbox, and a shaft. All of which make for an expensive, useless, lawn ornament without a tractor to power them.

I want you to find a tractor, almost any farm tractor will do. I want you to stand behind it. Look at the center of the seat's back and follow the imaginary center line of the tractor downward. Just south of the seat bottom and just north of the hitch (the thing at the bottom of the tractor with the hole in it used to tow things) you will notice a round, splined, shaft protruding from the tractor. This is the PTO (power take off) shaft. If you look at an implement, like our rotary cutter example, you will find that it has a long shaft beginning at the gear box and making its way towards the front of the cutter. At the end of this shaft is a hole (female) that matches up to the shaft that comes out of the tractor (male). When you put these two items together, turn on your tractor, and then activate the PTO, you suddenly have power going from your tractor to your implement! The implement can now do work! Think of an electric drill and a set of drill bits. The bits cannot do very much on their own, and while the drill makes noises and goes round and round, it doesn't do anything useful on its own. If you put the drill bit into the drill, you now have a too. By attaching your tractor to an implement, you now have a tool.

The beauty of the PTO system I just described is that it is 99.9% universal. It does not matter what brand of equipment you purchase, you will find this system in place. Having such a widely used system also allows you to mix and match implements with your tractor regardless of brand in the vast majority of cases.

Finally, it should be noted that there are two common PTO speeds: 540 RPM and 1000 RPM. Your implement's PTO speed and your tractor's PTO speed must match, so always ask before buying, "is this implement 540 or 1000?" Some tractors are equipped with both speeds. Most everyone familiar with equipment will be able to answer this question. The concept of "PTO Speed" sounds complicated, but it is not, the key is to know what PTO speed your tractor has, 540, 1000, or both, and buy implements that match.

Review: PTO = Power take off. A shaft on a tractor (male) that matches with a shaft on the implement (female) to draw power from the tractor to the implement. Ex) A rotary cutter (bush hog mower) needs power to go round and round. You achieve this by attaching the shaft on the tractor to the shaft on the mower. Make sure you only buy implements that match your tractor's PTO speed, 540 or 1000.

(example of a 1000 RPM PTO shaft. Compare to the earlier picture of a 540 PTO shaft. The splines on the 1000 RPM PTO are closer together and more numerous than those on the 540 RPM PTO)

(this rotary cutter is powered by a PTO. The black shaft pictured here is the female end that connects to the PTO shaft on the tractor)

Chapter Five: The Three Point What?

Another crucial aspect of getting your implements and tractor working together is the three-point hitch. The three-point hitch is the set of arms on the back of your tractor. There are two side arms and one top arm known as the "top link". They form a triangle. Each arm and the top link have holes at the end of them. The holes located at the end of the three-point hitch attach to corresponding pins on your implements and allows the implements to be lifted up and down via the tractor's hydraulic system. There is a lever on the side console of the tractor that controls this function. On most tractors you can even change the speed of how fast the implement goes up and down.

(three-point hitch: The two arms closest to the tires and the arm in the center pointing upward, known as the "top link", make up the three point hitch. Various implements attach themselves to the three-point hitch. The tractor operator can then raise and lower them)

(closeup view of one of the three point hitch links. This particular tractor has hitch links that slide out for easier hookup. The implement will connect to the round holes on the links)

When will you use this? The most common time I can think of is if you are doing grading with a rear blade, but there are hundreds of implements that are raised and lowered with the three-point hitch. Some items that attach to the three point hitch might also require PTO power such as three point hitch rotary cutters (known as "lift type" as opposed to "pull type" rotary cutters) and post hole diggers. Other items, like the aforementioned grading blades, do not require PTO power. If you look at the PTO shaft we discussed earlier, you will see that it is in the center of the three point hitch arms. This allows the three-point hitch and the PTO to be used in unison when the PTO required to run the implement.

WARNING: NEVER EVER EVER NEVER UNDER ANY CIRCUMSTANCES EVER should you hook a chain or rope to your three-point hitch in an attempt to tow or pull something out of the ground. You can use your tractors drawbar for this, but NEVER use the three-point hitch as it is too high up on the tractor's center of gravity and could cause you to flip over.

(here we have a neat two row corn planter that is raised and lowered by the three point hitch)

(This compact utility tractor has a spin spreader on the three point hitch. The PTO powers the spinning motion of the spreader)

Hydraulics: The Lifeblood of the Tractor

A very important, but often overlooked aspect of tractors among new buyers is the hydraulic system.   Most tractors have a series of outlets that are called any of the following: Selective control valves (SCV'S), hydraulic outlets, outlets, hydraulics, remotes, and hydraulic services to name a few. These valves are often located in the mid section of the tractor under the floorboard and or on the rear of the tractor behind the operator seat. The mid-mounted valves are typically used to control a front-end loader and the rear valves are used to operate hydraulic functions on rear implements. Small tractors usually have the option of two mid valves and two rear valves, though some sub compact models only offer one or no rear valves.

(rear mounted hydraulic valves: also known as "selective control valves" (SCV'S). These are used to power any implement that utilizes a hydraulic cylinder or pump)

(on implements that utilize hydraulics you will find rubber hoses with male tips that go into the SCV's on the tractor)

(the two black levers control the hydraulic valves. Each set of valves has a single control lever. Some fancier tractors utilize button and even touch screen hydraulic valve controls)

You will use the rear mounted valves to power such things as wood splitters, the angling cylinder on a blade for grading, wings on a batwing mower (larger applications), the ejection gate on a round baler, and any other implement function that requires hydraulic actuation.

The way in which hydraulics work are by forcing oil at high pressure through special rubber hoses and into metal cylinders that extend when filled with oil and retract when the pressure is let off. The hydraulic levers on your tractor are most likely located on the right hand console. If you have more than one set of hydraulics in the rear (two holes equal one valve, one hole provides pressure into the implement, while the other takes pressure away from the implement), the control and the hydraulic valves will most likely be marked with a roman numeral 1 and 2. While it is not uncommon for large farm tractors to have three, four, and even five sets of valves, this is not common, and rarely necessary in smaller applications. Most folks can get away with two mid valves for a loader and one or two rear valves for implement applications.

I highly recommend buying a tractor with two mid valves and at least one rear valve even if you do not think they will be of immediate use to you. Installing them later will be more expensive than ordering the tractor with the valves already installed. The older a tractor gets, the more expensive add on options like hydraulic valves become. Furthermore, if you decide owning a tractor is not for you or you wish to trade your tractor in, having mid and rear valves already installed makes your tractor far more valuable to both dealers and end users alike.

Front Loaders: A Real Day Lifter

Over the years as an equipment salesmen, many new tractor buyers came to me insisting all they needed was a tractor with a mower deck to cut grass. They saw no use for a loader. Though I always tried to convince them that a loader is the most versatile tool on the farm, most of them bought what they wanted, leaving the loader in my warehouse. Almost universally, six months to a year later, the customer that turned down my loader suggestion would call and ask me to sell them a loader for their tractor.

For those unfamiliar, a loader is the thing that goes on the front of the tractor, is often equipped with a scoop, known in industry circles as "a bucket", and allows the tractor to lift heavy objects like dirt, rocks, logs, bricks, dead pigs, lazy co-workers, you name it!

(the contraption with the scoop on the front is a loader. Here we see it in the down position)

(here we see the same loader in the up position…talk about a lift!)

The loader runs off of the tractor's hydraulic system. On new tractors, you will find most equipped with a joystick (like Pac Man) that controls the loader functions. Loaders have two basic functions that are operated by two separate hydraulic valves: Boom Up/Boom Down- this is the action that the long arms of the loader make to lift the boom and bucket into the air and bring it back down to the ground. Second, Bucket Roll-which is the action that causes the bucket to rock backwards when picking something up, and to rock forward when dumping. If this sounds complicated, it is not.

The bottom line is, if you are going to spend money on a tractor, BUY A LOADER.

"But it will just be in the way!" is what you're thinking.

My first retort to that suggestion would be that loaders are easy to take on and off of tractors these days compared to just twenty years ago. Most are freestanding units, meaning you don't need a special rack or hoist for removing or storing your loader. The loader just sits there on it's own two legs and bucket until you are ready to connect again.

Second, the things you'll be able to move with a loader will help you create all the space you need to justify storing a loader when you are not using it.

Loaders can also be fitted with different front-end attachment such as pallet forks (like a forklift), spears that pick up heavy round hay bales, and various types of buckets depending on the job you need to do. You can even put a snow blade on your loader!

Loaders help to eliminate wheelbarrows, armloads, handcarts, and dollies, along with the physical effort involved in using these tools in most situations.

In my experience, "go grab the loader tractor" is a common phrase around every working farm. Don't waste your back and your knees! Buy a loader when you purchase your tractor. You will not even realize how many things you can do with a loader until you are out there working on your farm, so if you are going to take the plunge, why not buy an extremely versatile tool with your tractor on day one?

Transmission Options: Just Like a Car, Only Totally different!

Just like your car, a tractor has a transmission. There are two basic transmission options in tractor transmissions: hydrostatic and gear drive.

In most of your compact utility tractors (20-55 PTO horsepower +/-), you will find dealers stocking hydrostatic transmissions. The reason for this is because the market has spoken. Most folks drive automatic cars and the tractor hydrostatic offers similar simplicity. With a hydrostatic, you have a high range (fast) and a low range (slow, but full of torque) and two pedals, one for forward and one for reverse. You put the tractor in the desired range, and then you push the pedal for the desired direction. The farther you push the pedal, the faster you will go.

From a learning standpoint, hydrostatics are very easy to operate. From a functionality standpoint, unless you are serious about doing farm work like hay baling, a hydrostatic is probably fine. Hydrostatic transmissions are far less problematic than they used to be and they make up the majority of new tractors under 65 horsepower sold on the market today.

(The two pedals are forward and reverse pedals for a common hydrostatic transmission)

Traditional farm tractors tend to be gear driven. A gear drive tractor is akin to a manual transmission car. You press a clutch and you pick your gear. The advantage to this in farm work is the ability to pick a ground speed and stay with it while doing a certain task.

Tractor manufacturers have made the job of shifting easier by building better clutch packages and adding a mechanism called a "reverser" to many models. The reverser allows the operator to toggle a lever (usually near the steering wheel), which allows the tractor to go from forward to reverse without engaging the clutch or touching the gear levers. This is an especially nice feature when you are doing work with a loader such as moving a pile of dirt or unloading pallets of feed from a truck.

(Transmission range and gear selectors. These will usually be orange. In this picture, they are the bright white knobs)

(Power reverser switch on the steering column. Some brands and models put them on the dash or on a side console. This allows you to switch from forward to reverse without switching gears or clutching in most cases. Notice the "N" on the column to the left of the steering wheel. That stands for "neutral". To engage forward gears, toggle forward. To engage reverse gears, toggle towards yourself. Pretty simple!)

If you are serious about farm work such as making hay or light tillage, my recommendation is a 12/12 or 16/16 reverser transmission. This simply means 12 or 16 forward gears with 12 or 16 reverse gears. There are cheaper options and there are more expensive options, but if you are springing for at least a 65 PTO horsepower tractor (ex: John Deere 5000 series), it is best to have this style of transmission from both an operational and resale standpoint

Visit a dealership and have them show you how the transmissions work and give them a try. Don't be afraid. Once you have the basic idea, tractor transmissions are easy to understand. As with all things, you can't learn until you try!

Tires? Yes, I want Tires. What's with all the options?

Tires are a very important consideration in the tractor buying process. The right tires can be the difference between a job done right and a job done UGLY.

There are three basic tractor tire options available on most tractor models. Agricultural tires (R1), turf tires (R3), and industrial tires (R4).

(R4 Industrial tread.)

(R3 Turf Tread)

(R1 Agriculture Tread)

The agricultural tire tread is what you see on farm tractors out in the fields. This tread is the most aggressive tread as it is typically used in heavy field applications where traction is critical in various conditions. An agricultural tire, which you may also hear referred to as an R1 tire, is great if you only plan to use your tractor in mucky barnyards, the woods, or in the fields. R1 tires are great for keeping tractors from getting stuck. They are not so great for keeping a lawn free of ruts.

That said, our next tire option is the turf tire, also referred to as an R3. A turf tire offers a smooth, even, tread, designed with manicured lawns in mind. If your main goal is heavy yard work or maintaining your own golf course, this is the tire you want, as it will not degrade your manicured turf grass. If you take these tires into a muddy pasture though, you may end up on your cell phone calling Farmer Fred to come pull you out with his big ag-tire tractor.

That brings us to our third choice, a tire option that offers the best of both worlds! The industrial tire, known in industry circles as an R4 and sometimes as a "bar" tire due to the way the tread looks. The R4 is what I like to call a "middle of the road" tire option.

R4 industrial tires offer a pronounced, but flat tread that can give your tractor better traction than a turf tire, but without the aggressiveness of agricultural tires. The R4 is a great tire for multi-use tractors among budget minded users. If you can afford a dedicated lawn and garden machine for your nice grass and a farm tractor for your dirty work, I recommend turf tires on the lawn unit and R1 agricultural tires on the farm unit. However, if you are like most of us and can only afford one machine for multiple tasks, the industrial tread is a great all purpose tire.

Agricultural Tires: Pros- Best traction in varying field conditions and helps maximize efficiency in field operations. Ag tires are also great for snow removal. Cons- aggressive tread is not good for lawns. Tread wears out quickly when used on hard surfaces such as cement and pavement.

Turf Tires: Pros- Will not hurt your grass. Cons- Not good in treacherous areas.

Industrial Tires: Pros- middle of the road tire between turf and ag. Won't destroy lawns. Won't wear out as fast on hard surfaces, great for concrete barnyards or loading docks. Offers enough traction for off-road conditions.

Cons- Simply is not an ag-tire. Simply is not a turf tire. In an ideal world, you would own one dedicated machine for manicured ground with turf tires, and one dedicated machine for farm-applications equipped with ag-tires. Barring the ability to buy two tractors, industrial tread tires are a good option.

Your Dealership: Not Just a Transaction, a Partnership

I cannot stress to you enough the importance of a good dealership in regards to your tractor and implement buying and owning experience.

I don't care if you are a master mechanic or a total know-nothing. You will need your dealership after the sale to remain a happy equipment owner. Sure, you can do your own maintenance, but most of you will not have the time or ability beyond basic oil and filter changes. If a bigger problem arises, will you have the tools, knowledge, and resources to make the fixes on your own? If you do have the ability to make these repairs, where will you buy your parts? If you need additional implements, where will you go?

My goal here is not to steer you towards any particular brand, I do however recommend trying to work with the dealership in your local area that has the best reputation for service. While it isn't always true, the dealership that has been in town for decades and is still doing a strong business among the rank and file farm community is most likely your best choice. If that business can maintain machines and customer service when big money is on the line (like crops that must come in NOW and narrow planting windows), they can surely cover you when your weekend warrior machines need work or you need parts.

Ask around. Who has the best reputation for quality? Which dealer will take calls from working farmers at midnight when a combine is broke down in a field? Who stays open the latest on Saturdays? Who has the least amount of turnover among mechanics and parts personnel? You will be amazed at how frankly your local farmers will answer these questions. Professional farmers and equipment dealers are business partners. The dealer with the reputation as the best partner to the farm community will most likely be your best bet as well.

Remember, any dealer can sell you a tractor and implements, but who will help you when you have a question on operation, adjustments, or need parts or service in a pinch? Start asking around before you start shopping!

You may want to take a Saturday morning and just visit dealerships within a certain radius of your farm. You don't have to talk to anyone at the dealership. Just make a list of local dealerships and go to browse around. Watch how people interact with customers. If you can, poke your head in the service area. Does it look busy? Busy is good. Does it look like a junkyard? This is usually bad. A Saturday spent just driving around and kicking tires with no intentions to buy could give you some good insights as to which dealers in your area you will want to deal with.

Dealing With Sales Representatives

Most of you have probably had both good and bad experiences with sales professionals. Buying a car is my least favorite experience. Buying a Sleep Number Bed® however was a dream! Buying a tractor falls somewhere in the middle and may vary from dealership to dealership.

Tractor salesmen and dealerships, assuming you have chosen one with a long and respectable track record in the ag-community, are typically not like a car dealership. Cars are a volume game, which leads to high-pressure tactics.

In contrast, tractor salesmen aren't just trying to sell a bunch of units. They are usually required by their employer to make positive margins and they want to solve specific customer problems. They won't play the "what payment can you afford game?" in most cases. A tractor sales representative may ask your budget at some point in the discussion, but their goal will be to get you in a tractor that fits your needs, one that won't be overkill and one that won't leave you wanting more. Tractor dealers also want to sell you all of the proper implements needed to do the jobs that you wish to do. The car ultimately only serves one tangible purpose: transportation. A tractor is a multi-use tool that comes in many different sizes and configurations, so there is more at stake for a tractor salesman if he doesn't get a customer matched with the tools to suit his or her needs. A Fortune 500 CEO and a mailroom clerk are both equally served by a Honda Civic if emotional needs such as "image" are thrown out. However, two farms that literally share a property line could be totally different in regards to the machinery they require.

If you are in a tractor dealership and it starts feeling like that really bad car dealership feeling you had a few years back, leave, it should never be that way in an equipment dealership (I have dealt with hundreds). There are plenty of farm equipment dealers that don't play like that. There is generally not the back and forth haggle process in the tractor business either. Dealers know the margins they need achieve in order to stay in business, they don't get huge manufacturer kickbacks, and they want satisfied, repeat customers. They also know they will be the ones selling you parts, service, and additional implements down the road.

Most sales reps in the tractor industry have a farm or construction background. They have literally run the equipment on their own farms or farms they have worked on. While their interest is to sell you equipment, most of them are great sources of first hand information on operation and maintenance. Ask them about their personal experiences and about their backgrounds. Like most people equipment sales reps will be anxious to share their personal stories and experiences with you. This also helps break any tension that there may be and can open up a positive dialogue about your wants and needs.

That said, when you walk into a farm dealership, leave those bad experiences you may have had in other businesses behind you. While it isn't unheard of, you will rarely find the sleazy, gum chewing, "WHATS IT GONNA TAKE TO GET YOU IN A (WHATEVER) TODAY???!!!" types.

Once you've found a dealership representative that you wish to work with, ask lots of questions. Like your farm neighbors, most sales reps in the agricultural industry are anxious to share their knowledge with open-minded folks who are seeking answers and willing to listen.

Admit you know nothing. This gets rid of any assumptions the sales rep may have had regarding your knowledge and will help him or her guide you to the right tractor and implements for your property.

Answer the sales reps questions honestly. They will almost surely ask you about property size and about the tasks you want to perform with your tractor. If you cannot answer these two questions, you are probably not ready to buy a tractor.

Ask the sales rep about his personal sales experiences. "What do most people in my situation buy? What would you buy if you were in my situation?" This may sound like a trap, but it is not. You will find that the vast majority of farm equipment sales reps are very honest and straightforward. Remember, a tractor in your case is a probably a one time purchase, he wants you to be satisfied.

The Best Time to Buy

There really is no best time to buy a tractor. Unlike cars, tractor models don't change every year. There won't be any "BIG MODEL CLOSEOUT ONE WEEK ONLY UNTIL THEY'RE GONE!" event at your local farm machinery dealership. However, in late winter and early spring, manufacturer's often give dealers incentives to sell to folks in the "lifestyle" or part time farmer market segments. If you're looking for a compact utility tractor, late winter and early spring will most likely offer the best cash discounts and or financing packages. This varies from manufacturer to manufacturer and prices will vary between dealers, but in my selling experience, the most attractive discounts and options for the utility tractor sector tend to start around February and run through May or June. Obviously economic circumstances will dictate this.

For example, as I write in late September of 2012, I have seen all of our "pre season" incentives for tractors between 25 and 100 horsepower extended through fall as my manufacturer has seen a slow down in units due to the economy.

The bottom line is there is no perfect time to buy. There are slightly better times (pre-usage season) and there may be economic circumstances that make for even better buying opportunities, but there is no absolute best time to buy. When to buy is truly a moving target based on your own timeframe and the decisions of the manufacturer.

While there may not be a best time to buy, there is a best time to shop. I recommend visiting your dealership in any of the following time frames: In the dead of winter or the heat of summer. These are the times of year that most sales representatives in the farm equipment business have the most time on their hands. Going shopping during the peak of planting, harvest, or year end tax season is fine, but you may have trouble finding a representative that can give you their full and undivided attention unless they only sell to the part time and lifestyle farm segment.

Rotary Cutters: When a lawnmower just won't cut it.

When purchasing your first tractor, you will obviously need to buy an implement or two to make your tractor useful. A common first purchase is a rotary cutter.

Lets face it some areas on our farms are just too tough to cut with a traditional lawnmower. Lawnmowers are designed to cut grass on a regular basis. This grass is found on maintained ground that is generally defined as "a lawn". Fields, pastures, banks, fallow fields, and open spaces that don't need to be cut as often require something with a little more muscle.

Enter the rotary cutter, an implement that goes behind the tractor, is driven by the PTO, and utilizes a heavy blade that cuts through grass, brush, and debris. Some rotary cutters are rated to handle brush up to 4 inches in diameter! The one you buy will most likely be rated for up to two inches of brush diameter, which is more than enough in most cases.

The key to selecting a rotary cutter is tractor horsepower, tractor width, and three point hitch category (limited 1, 1, 2, 3). Larger width rotary cutters also come in pull type varieties that require less horsepower and do not utilize the three-point hitch. Pull type rotary cutters are nice in open spaces, but can become cumbersome in tighter areas.

When selecting a rotary cutter, make sure the width covers your tire tracks. Otherwise you will find mowing to be quite tedious.

Now that you know the basics of rotary cutters, you can stop destroying your lawn mower out in your rough fields!

Possible uses for rotary cutters: Along farm lanes, cutting down gardens in late fall, clearing brush from fence rows, mowing pastures, trimming right of ways, chopping up cornstalks, cutting down patches of invasive weeds.

* Rotary cutters are often referred to as "bush hogs" which is actually a brand name and "brush hogs" which is a variation of that brand name.

(Six foot three-point hitch rotary cutter. The long shaft in the center connects to your
tractor's PTO, while the triangular apparatus connects to the three-point hitch)

(For larger properties that need lots of mowing, consider a batwing style rotary cutter. The wings fold up and down with the tractor's hydraulics allowing for easier transport and maximum mowing width)

Snow Removal: Blade, Blower, or Both?

In many parts of the country, snow removal is a regular, often hated, but necessary chore in the colder months. Seeing as how you have bought or are about to buy a tractor that will most likely cost a minimum of $10,000, it sure would be nice to never have to touch a snow shovel again.

The two primary snow removal implements are snow blowers and snow blades. Both implements come in front and rear mounted formats and some utilize hydraulics for angling.

Depending on where you live, along with how much and what type of snowfall your area tends to get, will determine which snow removal strategy is best for you. If you live in a region that gets small dustings and the occasional "wintry mix" a blade is probably all you will need. A front blade mounts to the front of the tractor (like you see on trucks) and has a standard UP/DOWN function with optional hydraulic angling kits that allow you to control the angle of the blade from the seat of the tractor.

There are also blades available that mount to the three-point hitch. Both types of blades are effective, though the front mounts tend to be more expensive and require extra brackets in order to take the blade on and off. Rear blades simply mount to the three-point hitch. You will also need extra ballast, usually in the form wheel weights, if you are using a front blade. This is not as critical with a rear blade.

If you live in an area that gets heavy snowfall a snow blower is the way to go. The great thing about a snow blower is that you can blow the snow out of your way, where a blade may leave you with large, frozen piles, that take weeks to melt while more snow piles on top and in your driveway. The disadvantages to snow blowers are the fact that you have more moving parts (maintenance) and they need a lot of power to work effectively.

Depending on the size of the tractor you buy, a rear three-point hitch mounted snow blower may be the best option. Front mounted snow blowers, while nice in that you don't have to look behind you while working, are expensive. Unfortunately on smaller compact tractors, the rear mounted blowers simply require to much power to be effective, so a dealer will most likely recommend a front mount or a blade.

Regardless of what you use to remove snow, it is critical that the implement at least cover the width of the tractors wheel tracks (the width from the outside edge of one rear tire to the outside edge of the other rear tire). Otherwise, you will find yourself compacting more snow than you are moving. Moving snow becomes a long tedious process at that point.

I highly recommend rear mounted blowers whenever possible. Having worked with all kinds of snow removal tools in various conditions, these units do an amazing job when matched with the proper tractor. Other things to consider are hydraulic angling for blades (assuming your tractor has an additional hydraulic outlet) and hydraulic chute control for snow blowers. There is nothing worse than having to get on and off of your tractor in icy cold conditions to pull pins and levers to adjust a snow removal tools angle or chute direction.

If you are new to the area, talk to local farmers about what they use to move snow. Dairy farmers have to keep lanes cleared at all times so the milk trucks can pick up their milk, so dairy farmers are a good local resource on snow removal. The sales representative at your local dealership can also give you his personal recommendation based on what most people in the area buy and based on what he or she may use on their own property.

(Rear mounted blade for snow removal or grading. Attaches to the tractor's three-point hitch. Some blades utilize a hydraulic cylinder to angle the blade. This particular blade's angle is manually adjusted by pulling the pin located at the swivel point)

(Rear mounted snow blowers come in widths as small as four feet wide and as large as you can imagine. They are a great way to move snow but they require quite a bit of power)

Cash or Finance?

Lets face it, most of us need to borrow money when we purchase big ticket items. Why would we treat buying a tractor or other equipment any differently, especially in this age of super low rates?

Tractor dealers and manufacturers generally work their pricing like this:

You can buy a tractor and take a "cash" bonus. This means X dollars is taken off of the price of the tractor in lieu of using super low rate programs.

This doesn't mean you can't finance the tractor, but you most likely won't be able to get a special rate like 0% for 60 months.

If you want a super special rate like 0% for 60 months, you give up the "cash" discount.

Don't let the term "cash discount" (also referred to as a "retail bonus" sometimes) confuse you. You can still, in most cases, use dealer/manufacturer financing, but the interest rate will be based on your credit score and a standard rate tier alone and not some manufacturer sponsored special program such as 0% for 60 months.

Example: Tractor $20,000 -$1000 retail/cash bonus = $19000 financed balance (forget down payments for now, you may or may not need one).

You find out you qualify for the manufacturer's top standard rate of 4% interest.

If you take a 60-month loan at the 4% rate, your payment will be $349.91. The total interest cost over the life of the loan will be $1994.87, bringing the total tractor cost to $20,994.87.

Lets say the dealer says, "we have 0% for sixty months, but we'll lose the retail bonus of $1000 if you opt for that low rate program."

You would then finance $20,000 at 0%. Your payment would be $333.33 and no interest, bringing the tractor's total cost to $20,333.33.

In this scenario, you would actually save money by giving up the retail bonus and taking the 0% interest, assuming you need to finance. If you can actually pay with cash, you would cut the dealer a check and you would pay no interest and maximize the benefits of the cash discount/retail bonus.

This scenario will not always play out the same way, so you MUST RUN YOUR OWN NUMBERS!! Get the price both ways.

To avoid confusion, ask for both prices using this phraseology,

"cash price with standard financing and best price with special financing"

Ask the sales representative to write down the terms and the rates as well. Then you can go to a site like www.bankrate.com or pretty much any "simple loan calculator" site and determine what the best deal is.

The salesman cannot and should not make this decision for you. You need to ask yourself the following:

1) is getting the lowest monthly payment possible most important to me?

2) do I care how much I ultimately pay in interest over the course of this loan?

3) I have the money to pay for the tractor outright, do I want to use it?

Every individual situation is different. The key is to do your homework before making a decision. If your property is classified as a farm and you will be filing a Schedule F with the IRS, you may want to sit down with a tax accountant before making equipment purchases so they can determine the best purchasing strategy for you based on your income tax situation. Again, there is no cookie cutter way to determine which method of payment is best. Even for full time farmers, each year and each purchase will be based on a different set of financial circumstances. Know yours.

The Ever Crucial Backhoe (or not)

In my years of selling equipment, the most common perceived needs among new customers were, tractor, loader, and mower in that order. The one thing that got added to that list of needs quite a bit was a backhoe. A backhoe is an arm like contraption that sits on the back of the tractor and is run by hydraulics. They usually have their own little seat and the operator moves to the seat to run the backhoe. Backhoes are typically used for digging trenches.

My typical response to would be backhoe buyers was as follows, "unless you plan to go into the landscape contracting business or the plumbing business, forget the backhoe and just rent one when you need it." Believe me, at a minimum of $5000 a piece for a backhoe on a sub compact (more $$ depending on the tractor size), I would have loved to have sold backhoes to every Tom, Dick, and Henrietta property owner that came along. Unfortunately, this is what happens a few years after I sell someone a backhoe:

They try to use it.

They find that their tractor is way to small to do the backhoe job they want to do and that using a backhoe actually takes a lot of practice and patience, which translates to time, which is something weekend warriors usually don't have a lot of. They end up paying someone to do it or they rent a bigger machine to meet their digging needs. Then the tractor owner sticks their $5000 backhoe (minimum of $5000 remember) in the corner of the shed and it sits for a few years. Suddenly they decide it is taking up too much space and they call me,

"Hey, Mr. Tractor Guide Guy, ummm, remember that backhoe you sold me? Yeah, I've had it for two years now, used it once, anyway will you buy it back?"

To which the answer is almost always, "No thanks."

The next question is, "can I trade it on something else?"

To which the answer is "sure".

Next I give them a trade in value that is so low they get angry with me and never want to do business again.

So why won't a friendly neighborhood salesmen trade in a backhoe for top dollar?

A) Most backhoes are designed to fit specific tractors. Therefore, they can only be sold to someone with that specific tractor.

B) Backhoes take up space on our dealership lots and they are hard to market by themselves unless a person has the exact same tractor as you.

C) Tractor configurations change so often that sometimes something as complex as a backhoe will not retrofit to newer units without massive fabrication.

While backhoes are an invaluable tool in the hands of a landscaper, an irrigation contractor, or a plumber, given the cost, the space they take up, and the time it takes to become a proficient operator, I simply don't recommend them for the average entry-level tractor buyer. The purchase does not make economic sense unless you have money to lay in the corner of your barn, collecting dust, and never being used. I recommend renting a backhoe on an as needed basis. Better yet, find a neighbor with a backhoe who has spent 10,000 hours learning how to use it properly and pay him or her to the job.

How Will I Transport My Tractor if I need Service?

The answer to this question is simple: you probably won't have to. Of course, your tractor and any number of implements will need service at some point. Unfortunately, not all problems can be fixed by a service technician right on the farm, so at some point, your tractor may have to go back to the dealership for service. Many folks think,

"Man, I'll have to buy a trailer to bring this tractor back to the shop should it break down."

That is not the case. Most dealers have a small fleet of large trucks and trailers to handle transporting machinery to and from customers. Any dealer that doesn't offer trucking probably won't be a dealer much longer and you should not deal with them. Obviously, you will have to pay for trucking in many instances, but in the vast majority of cases, it is cheaper and more efficient to pay a dealer for hauling than it is to invest in a big truck and trailer. There is no need to worry about trucking if you buy from a reputable dealer. Most of them have to be in the transportation business if they wish to be in the farm equipment business.

Self Service or Not?

If you can change any fluid in your car, you can do the same thing in a tractor. Every tractor has engine oil and a filter. Other common do it yourself items include fuel filters, air filters, transmission fluid, hydraulic fluid, and coolants. If you have experience changing these things in passenger automobiles, you will find little to no difference when you try to service your tractor.

The main thing to do before attempting to service anything on your own is to read your operator's manual. It will give a breakdown of when you need to have certain things serviced. The operator's manual will also tell you the filters you will need and the proper fluid types to use in your specific tractor. Many farm machinery dealers charge between $50 and $75 an hour for labor, so there is some savings to be had by doing your own fluid and filter changes. Repairs you make on your own above and beyond the most basic are completely up to you, though you should use care in that you don't void your warranty and that you have the proper tools at your disposal.

Now that you know the basics about tractors, common implements, and the buying process, I suggest you spend a Saturday kicking tires. You should also start talking to your neighbors about what they use and recommend. I could write 1000 more pages and not tell you everything there is to know about farm equipment, but you should now have the basic knowledge to go into a dealership and begin asking educated questions and answering the sales representative's questions.

Going back to the first page of this manual, you need to remember two things. 1) You are not the first person to do this. 2) There are people out there who are willing and able to help you learn. Don't be afraid of things you are about to learn and the work you are about to pursue. Embrace it.

Happy farming!

www. ngramcontent.com/pod-product-compliance
Lightning Source LLC
Chambersburg PA
CBHW081154040426
42445CB00015B/1877